FLORA OF TROPICAL EAST AFRICA

TRAPACEAE

J. P. M. Brenan

Aquatic floating herbs. Leaves alternate, floating, rosetted, only present at the upper nodes ; stipules small, scarious, cleft to base and thus apparently more than two per leaf ; petiole spongy and more or less inflated about the middle ; lamina rhombic to deltoid. Adventitious roots (?) submerged, paired but not opposite, one from either side of leaf-scar, chlorophyllose and thus leaf-like, pinnatisect into many filiform segments. Flowers solitary from upper axils, pedunculate, hermaphrodite, regular. Sepals, petals and stamens 4, latter perigynous. Petals white. Ovary half-inferior, bilocular ; ovules pendulous, one per loculus. Fruit a one-seeded, top-shaped drupe ; pericarp soon disappearing ; endocarp very hard, variously 2–4-horned, the horns derived from the persistent sepals.

Clearly related to and probably derived from *Onagraceae*, but so distinct in its morphology that it is better as a separate family. The germination is very remarkable, one cotyledon only being exposed, the other remaining within the pyrene (endocarp)—see Hegi, Ill. Fl. Mitteleuropa 5 (2) : 889 (1925) for illustrations.

I do not feel altogether satisfied with the interpretation of the submerged pinnatisect organs as adventitious roots, since they are so regular in their number and position, but have so far not found any published anatomical evidence about this.

The family has only the following genus.

1. **TRAPA**

L., Gen. Pl., ed. 5 : 56 (1754)

Characters of the family.

1. **T. natans** *L.*, Sp. Pl. 120 (1753) ; Hegi, Ill. Fl. Mitteleuropa 5 (2) : 884 (1925). Type : presumably a European plant

Annual. Leaves with lamina often broader than long, 1·5–4·5 cm. long, 1·5–7 cm. wide, toothed in upper part, glabrous above, beneath varying from almost glabrous to densely hairy all over. Sepals about 4–7 mm. long. Petals obovate-oblong to obovate, about 8–16 mm. long. Fruit very variably 2–4-horned.

var. **africana** *Brenan* in K.B. 1953 : 171 (1953). Type : Uganda, *Snowden* 1832 (K, holo. !, BM, iso. !, EA, iso. !)

Pyrene (i.e. endocarp) 4·5–6 cm. across in all, 4-horned, horns slender, straight or very slightly curved, 1·5–2 cm. long and 2–3 mm. wide near base, sharp at point and reflexedly barbed for a little way below it, two of them deflexed and two nearly horizontal. Fig. 1.

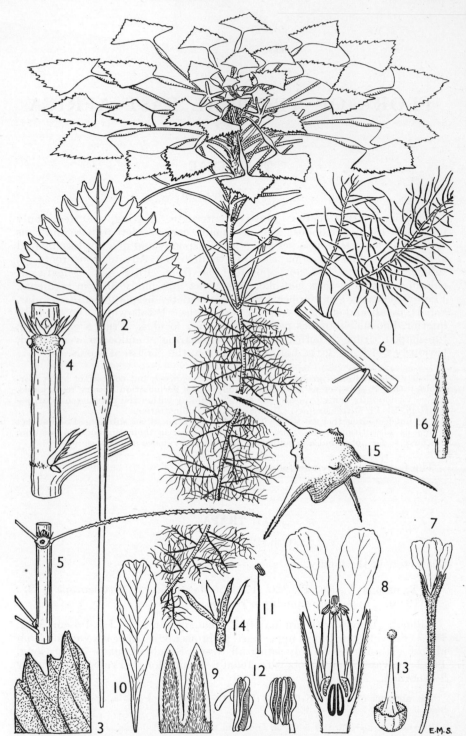

FIG. 1. *TRAPA NATANS* var. *AFRICANA*—1, part of plant showing floating leaves and submerged adventitious roots, × 1/3 ; 2, floating leaf, × 2/3 ; 3, marginal part of leaf, lower side, natural size ; 4, nodes showing stipules, × 3 ; 5, nodes showing first stage of adventitious roots, × 1½ ; 6, nodes showing fully developed adventitious roots, × 1½ ; 7, flower, natural size ; 8, flower cut longitudinally, × 3 ; 9, two sepals, × 3 ; 10, petal, × 3 ; 11, stamen, × 3 ; 12, anther, two views, × 12 ; 13, ovary and disc, × 3 ; 14, ovary and calyx enlarging after anthesis, natural size ; 15, endocarp of fruit, natural size ; 16, apex of one of the horns of the fruit, × 3.

UGANDA. Mengo District : Lake Victoria, Entebbe, Nov. 1930, *Snowden* 1832 !; same locality, Mar. 1932, *Tothill* 1565 !

TANGANYIKA. Lake Province : Ukerewe, *Conrads* 5894 !

DISTR. **U4** ; **T1** ; apparently endemic to Lake Victoria, but uncertain in absence of fruiting specimens from other places ; a sterile specimen from Uganda (? **U1** or 2), Nile 2° N. of the Equator, 10 Nov. 1862, *Speke & Grant* 515 ! may well be var. *africana*

HAB. Floating in quiet water of lake, 1136 m.

SYN. *Trapa africana* Flerow in Bull. Jard. Bot. Républ. Russe 24 : 45 (1925), *nomen nudum* ; a probable synonym

Trapa natans L. var. *africana* (Flerow) Gams in Pflanzenareale 1 : map 25 (1927), *nomen nulum* ; a probable synonym

var. **bispinosa** (*Roxb.*) *Makino* in Iinuma, Somoku-Dzusetzu (Iconography . . . Plants . . . Japan), ed. 3, 1 : 137 (1907). Type : India, *Roxburgh* (BM, lecto. !)

Pyrene (i.e. endocarp) about 3–5 cm. across in all, 2-horned, horns arising from the upper angles, erecto-patent to arcuate-ascending or almost horizontal, straight or somewhat curved, conical, or attenuate above, about 1–1·8 cm. long and about 4–7 mm. wide near base, sharp at point and reflexedly barbed for a little way below it.

TANGANYIKA. Kigoma District : Lake Tanganyika, mouth of Malagarasi R., Jan. 1905, *Cunnington* 43 !

DISTR. **T4** ; Portuguese Guinea, Anglo-Egyptian Sudan, Portuguese East Africa, Nyasaland and Angola ; also from India to China and Japan

HAB. Floating in lake, 773 m.

SYN. *Trapa bispinosa* Roxb., Fl. Ind. 1 : 449 (1820)

The var. *bispinosa* has also been collected in Nyasaland by Lake Nyasa only about 20 km. from the Tanganyika boundary (*Cunnington* 13, at BM !), so that it almost certainly occurs in that part of our area also. Its apparent absence from Lake Victoria and replacement there by var. *africana* is noteworthy. Fruiting material of *Trapa* from tropical Africa is very sparse, so that var. *bispinosa* will doubtless be found to occur in other African territories than those given above.

DISTR. (for aggregate species). A very wide discontinuous range through Europe, Asia and Africa (north Africa to Natal) ; naturalized in northern America and Australia ; more widespread in Tertiary times than to-day

Some authors have recognized many distinct species of *Trapa*, entirely based on the form and size of the pyrene of the fruit. Believing that a species should have more than one character, and recalling the variability of the pyrene of *T. natans* even in Europe, whence both 2- and 4-horned variants are on record, I prefer to follow the wide view of Gams as expressed in Die Pflanzenareale, with the reservation that I do not follow his reasons for subordinating the 4-horned var. *africana* to the 2-horned *T. natans* L. subsp. *bispinosa* (Roxb.) Gams.

INDEX TO TRAPACEAE